Housing
232

地下生活真快乐

Happy to Live Underground

Gunter Pauli

[比] 冈特·鲍利 著

[哥伦] 凯瑟琳娜·巴赫 绘

闫宇昕 译

上海远东出版社

丛书编委会

主　任：贾　峰

副主任：何家振　闫世东　郑立明

委　员：李原原　祝真旭　牛玲娟　梁雅丽　任泽林

　　　　王　岢　陈　卫　郑循如　吴建民　彭　勇

　　　　王梦雨　戴　虹　靳增江　孟　蝶　崔晓晓

特别感谢以下热心人士对童书工作的支持：

匡志强　方　芳　宋小华　解　东　厉　云　李　婧

刘　丹　熊彩虹　罗淑怡　旷　婉　杨　荣　刘学振

何圣霖　王必斗　潘林平　熊志强　廖清州　谭燕宁

王　征　白　纯　张林霞　寿颖慧　罗　佳　傅　俊

胡海朋　白永喆　韦小宏　李　杰　欧　亮

目录

Contents

一只树袋熊嗅到燃烧的气味，开始向安全的地方移动。一只巨蜥看到树袋熊全家缓慢移动的步伐，便说道：

"如果你们不想被烧焦，动作最好快一点！火灾正变得越来越严重！"

"呃，全球气温不断上升，气温越高，整个地区受灾的风险就越大。"

A koala smells fire and starts moving towards safety. A goanna watches him and his family walking away slowly, at a koala's pace, and says:

"You had better hurry if you don't all want to get fried! These fires! They are just getting worse and worse, day after day, year after year."

"Well, temperatures are rising all over the world, and the hotter it gets, the more the risk that a whole region will be destroyed."

一只树袋熊嗅到燃烧的气味，开始向安全的地方移动……

A koala smells fire and starts moving to safety …

······在树顶避难。

... refuge in the treetops.

"而且你们树袋熊有在树顶避难的坏习惯。你们难道没有意识到在那里全家都会被烤焦吗？"

"你想让我怎么做？我根本没法跑，因为我的四肢不适合奔跑。"

"能跑多快其实差别不大，因为你跑不过野火。甚至袋鼠都会大批地死去。求生的唯一机会是挖很深的洞。"

"And you koalas have a bad habit of seeking refuge in the treetops. Don't you realise that up there your whole family will get roasted?"

"What would you have me do? I simply cannot run, my limbs are not made for that."

"How fast one can move makes no difference, as you cannot outrun a wildfire. Even the kangaroos die by the hundreds. The only chance to survive is to dig deep into the ground."

"是的，你们巨蜥藏在洞穴中让火从上面掠过的做法是明智的。之后，你可以享用额外的晚餐，包括烤熟的蛇、蜥蜴和青蛙。"

"哦，我难道没有权力享用一顿额外的晚餐吗？浪费掉动物们的尸体真是太可惜了。起风时，氧气助长了火势，我的晚餐很快就准备好了。"

"Yes, you goannas are wise to hide in your burrows and let the fire pass over you. And afterwards, you enjoy an extra dinner, of fried snake, lizard and frog."

"Well, don't I have the right to an extra meal? It would be a pity to let their sacrifice go to waste. When the wind blows, the oxygen fuels the fire, and my dinner is ready in no time."

......你们巨蜥是明智的......

... you goannas are wise ...

火龙卷

Fire tornadoes

"我很怕那些把燃烧的残骸抛到几英里外的火龙卷。这些残骸甚至落在树顶——我避难的地方。然后树从上往下烧着了！"

"我很担心游隼捡起燃烧的树枝并扔在其他地方引发火灾。当生物从火中逃离，游隼能轻松地捕到猎物。"

"I am scared of these fire whirls, these fire tornadoes, that throw burning debris around for miles. It even ends up in the treetops, where I try to escape to safety. Then trees burn from the top down!"

"I am more worried about the peregrine falcons picking up burning twigs and dropping these to start fires elsewhere. When creatures flee from the fire, they have an easy meal."

"一切始于热浪和干旱，它们使所有植被都枯萎了。气候变化将继续给我们所处区域造成巨大打击，以至于很快你在任何地方，甚至在你的洞穴中，都不再安全。"

"你知道桉树也是问题的一部分吗？它会产生易燃的油，这样一来它就可以在自然竞争中打败那些不那么耐火的树种。"

"也许我们可以向非洲学习如何应对火灾。"

"And it all starts with heat waves and droughts that dry out all the vegetation. How sad that climate change will continue to ravage our country, so much so that soon you won't be safe anywhere, not even in your burrows, Goanna."

"Did you know the eucalyptus tree is part of the problem? It produces flammable oils to eliminate competition with other trees that are less fire-tolerant."

"Perhaps we can learn from Africa how to deal with fire."

桉树会产生易燃的油……

Eucalyptus trees produce flammable oils ...

树木相距甚远。

Trees are widely spaced.

"向非洲学习？" 巨蜥问道："那里难道不是贫穷的大陆吗？"

"非洲的稀树草原有值得我们学习的地方。"

"你是对的，每个人都有值得学习的地方。那些一生都过着舒适生活的人，其实可以向那些火灾中的幸存者学习。"

"你看，在稀树草原，树木相距甚远，所以土壤光照充足，草类植物可以大量生长。也许草看起来很干，但是一场充沛的降雨足以使不同的草产生比森林更多的植被。"

"Learn from Africa?" Goanna asks, "Isn't that the continent of poverty?"

"Their savannah may have something to teach us."

"You are right, everyone has lessons to offer and we, who have lived in comfort all our lives, could learn from those who have successfully survived the hardship of fire."

"You see, in the savanna, trees are widely spaced, so plenty of light reaches the soil, allowing for the grass to grow abundantly. The grass may look dry, but one good rain shower is enough for the different grasses to produce more vegetation than forests do."

"是的，那些植被为许多动物提供了牧草。非洲稀树草原容纳了世界上最后的大型哺乳动物群落，从食草动物到吃嫩叶的动物，以及捕食它们的大型猫科动物。"巨蜥说道。

　　"然后是不是在这里，在稀树草原上，人类首次开始直立行走？"

　　"这又与火灾有什么关系？"

"Yes, and that provides grazing for so many animals. The African savannah houses the last great mammal communities of Earth, from grazers to browsers, to the big cats that prey on them," Goanna adds.

"And was it not here, on the savannah, that people first started walking on two legs?"

"Now what does that have to do with fire?"

世界上最后的大型哺乳动物群落……

Last great mammal communities of Earth ...

一整棵树在地下？

A whole tree, under the ground?

"由于大火，稀树草原面积变大而森林面积缩减。早期人类移居到森林外面的草地，而一些树种则进化出了厚实隔热的软木状树皮来应对火灾的频繁威胁。"

"这些又能如何帮到我们呢？这里的树又无法很快改变树皮，现在它们做得到吗？"

"哦，其他的树为了生存转移到地下生活了。也许我们应该从那些树身上学习。"

"树，在地洞中生存？我知道树为了稳定性以及寻找水源和营养而在土壤中扎根。但是一整棵树在地下，这怎么可能？"

"Through fire, the savannah extended and the forests shrunk. Early humans moved out of the forests and into the grasslands. And also, some trees learnt how to cope with the constant threat of fire, by developing a thick, insulating, cork-like bark."

"How does that help us here? Our trees cannot quickly change their bark, now can they?"

"Well, other trees moved underground, to live there. Perhaps we stand to learn something from them."

"Trees, living in a burrow? I know trees have roots buried in the soil, for stability, and finding water and nutrients. But a whole tree, under the ground?"

"没错！这里有些树长得很大，直径可达10米，有着伸出地表并暴露在阳光下的小嫩枝。这些嫩枝非常小以至于如果偶尔遇到火灾，也不会有大问题。它们生长非常迅速。"

"所以，人们应该设计地下'大厦'来躲避地表灼热的熊熊烈火而不是建造高大的摩天大楼！好吧，安全确实比后悔更好。"

……这仅仅是开始！……

"Yip! Here some grow very large, as much as ten meters across, with tiny shoots on the surface that are exposed to the sun. These shoots are so small that it makes little difference if these get burnt in an occasional wildfire. They can quickly regrow."

"So, instead of building huge skyscrapers, people should design underground 'scrapers' to find safety from scorching fires raging above ground! Well, it definitely is better to be safe than sorry!"

... AND IT HAS ONLY JUST BEGUN!...

······这仅仅是开始！······

... AND IT HAS ONLY JUST BEGUN! ...

Did You Know?

你知道吗？

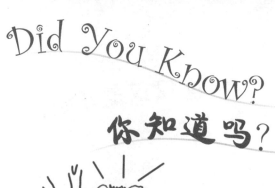

2020 年，澳大利亚南部海岸附近的袋鼠岛大火烧死了约 25 000 只树袋熊。保险起见，人们把岛上的树袋熊和大陆上的树袋熊分隔开。树袋熊运动缓慢，面临危险时会移动到树梢。

In 2020 some 25,000 koalas died in fires on Kangaroo Island, off Australia's southern coast. They were kept separate to those on the mainland as insurance for the species' future. They are slow-moving, and when in danger they move to the treetops.

仅在一个季节，丛林大火就向大气释放了约 3.5 亿吨二氧化碳。这相当于澳大利亚每年平均二氧化碳排放量的三分之二。重新吸收一年中排放的碳可能需要 100 年。

In one season alone, bushfires released about 350 million tons of CO_2 into the atmosphere. That is as much as two-thirds of Australia's average annual CO_2 emissions. It may take 100 years to re-absorb the carbon that has been emitted in one year.

传统的刀耕火种的生产方式，会砍伐树木、焚烧农田以促进草类生长。灌木因此成为环境中的主要植物，进而增加火灾的风险。

Burning of farmland to let grasses grow faster,cutting big trees leads to shrubs to dominate and this leads to risk of fire.

印度洋的暖流流向非洲。结果，寒流涌向澳大利亚沿海，使得海湾地区持续低气压并减少海水蒸发。这样一来，澳大利亚的降水来源被切断了。

Warmer waters from the Indian Ocean flow to Africa. As a result, cool ocean water wells up off the coast of Australia. This cool ocean water keeps atmospheric depressions at bay and results in less evaporation. Consequently Australia is cut off from the supply of rain.

Flames that reach up to 900 degrees Celsius in wildfires inevitably kill and injure many organisms in their path. A number of plants have adapted to re-sprout if they are damaged in a blaze. These re-sprouting plants include several Eucalyptus species.

野火中高达 900 摄氏度的火焰必然会杀死并伤害许多生物。如果在大火中受损，许多植物都可以重新萌芽。这些重新发芽的植物包括几种桉树。

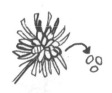

Some plants have fire-activated seeds. The lodgepole pine, Eucalyptus, and Banksia, have seeds completely sealed with a resin. These cones can only open to release their seeds after the heat of a fire has physically melted the resin.

一些植物的种子是遇火释放的。黑松、桉树和山龙眼的种子被树脂包裹。只有用火加热使树脂融化后，球果才能张开并放出种子。

依靠树皮中的隔热层，植物可以在野火中幸存下来。落叶松和巨杉有非常厚的阻燃性树皮，它们可以承受燃烧并且保证树的重要组织不受任何损害。

Plants survive wildfires due to a layer of thermal insulation in their bark. Larches and giant sequoias have incredibly thick fire-retardant bark that can be burned without the tree sustaining any damage to their vital tissues.

非洲地下森林的植物被称为地生茎植物，它们是通过在地下生存来预防火灾的灌木丛和树木，只有很少的芽和叶在地面上可见。南美洲的塞拉多也出现了同样适应火灾的地下植物。

Africa's underground forests called geoxyles are bushes and trees that are protected against fire by living underground, with only a few shoots and leaves visible above ground. The same fire-adapted underground plants emerged in the South American Cerrado.

Would you be prepared to live underground to be safe?

你愿意为了安全起见住在地下吗？

Is any member of your family a volunteer member of the fire brigade?

你的家庭成员中有消防队的志愿者吗？

Is a bird spreading fire an arsonist? Or is the bird just looking for food?

放火的鸟是纵火犯吗？ 还是说那只鸟只是在寻找食物？

Can we learn from the poor, or only from the rich and smart?

我们可以向穷人学习吗？还是只能向富人和聪明人学习？

Do you have a fire protection plan in place? When a fire breaks out people panic and, unable to reason clearly. Think of the ways you can ensure that everyone remains calm. Draw up a status report on your home, by first listing suggestions of how fire risk can be reduced.Share your report with your friends and family members. By training them in fire safety, you could save lives. One of the tricks we can all learn is to stay close to the ground where there are no toxic fumes, and cover our mouth and nose with wet handkerchiefs.

你有防火预案吗？火灾发生时，人们会恐慌并且无法清晰地思考。想想有哪些办法能使每个人都保持冷静。首先列出关于减少火灾风险的建议，并拟定房屋状况报告。与你的朋友和家人分享你的报告。通过消防安全培训，你可以挽救很多生命。所有人都能学会的一个技巧是，逃生时要贴近没有毒雾的地面，并用湿手帕遮住口鼻。

学科知识
Academic Knowledge

生物学	稀树草原不是贫瘠的生态系统，在雨季，该生态系统下长出的植被比森林更多；地生茎灌木或地下树木；地下树种有200种；木质块茎可以帮助植物抵御火灾；细菌相对于真菌具有更高的耐热性；通过C4途径固碳的植物都是被子植物；地下森林的年龄大约为13 000年。
化 学	与C3光合作用相比，C4光合作用在二氧化碳浓度较低时更高效；因为有灰烬作为肥料，像澳大利亚草树这样的植物在火灾过后会大量开花；温度升高时碳酸钙分解为氧化钙，土壤的pH值会因此升高；圣诞树上被喷上了阻燃剂；合成阻燃剂释放二噁英和呋喃。
物 理	稀树草原受益于野火，大火过后草再次发芽；软木状的树皮可作为防火材料；高大的树冠将叶子和重要的生长组织置于高过大多数火焰的高度，从而减轻野火的伤害，并为树袋熊提供庇护所；大火通过氧化、挥发、侵蚀和过滤使土壤养分发生变化。
工程学	探测、扑救和减轻火灾的技术。
经济学	纵火最常见的原因是为了利益（保险赔偿金）；当人们用钱抵偿死亡的风险时，成本效益分析理论在消防系统中得到实际应用。
伦理学	人们将大规模的野火归咎于灭火不当并继续在已适应了火灾的生态系统扩建居住区；阻燃剂是有毒的，这迫使人们在安全性和癌症之间做出选择；我们准备好向穷人学习吗？
历 史	稀树草原大约在800万年前开始在地球上扩张；草地代替了原始森林；詹姆斯·库克把澳大利亚称为"烟雾笼罩的大陆"；C4草类在渐新世统治了稀树草原，并在更新世冰期达到了数量的顶峰；希腊神话中浴火重生的"菲尼克斯"类似于中国的凤凰和朱雀，以及日本的凤凰。
地 理	稀树草原树木间距大，光照充足，草类因此占据支配地位；稀树草原和针叶林等生态系统会随着大火进化，从而促进栖息地的恢复；火灾是自然灾害，类似于洪水和暴风雨，驱使物种和生态系统不断进化。
数 学	消防安全定量风险评估。
生活方式	使用斯莫基熊来进行防火宣传，教育公众关于意外人为火灾的危险；在一些国家，消防队主要由志愿者组成。
社会学	过去人们常焚烧植被以方便狩猎，或以此促进某些植物生长。
心理学	纵火犯的典型特征是智商低，其行为常常受难以发泄的愤怒情绪支配，90%的纵火犯是男性，一半是未成年人；火灾疏散时的恐慌行为。
系统论	现代人类用火烧掉大面积的森林，促使稀树草原扩张；北美短叶松对于濒临灭绝的科特兰莺非常重要，这些鸟只会在幼松上筑巢，而这些树只能在最近被烧毁的森林中生长。

情感智慧
Emotional Intelligence

巨　蜥

巨蜥担心树袋熊能否活下来。他发现树袋熊移动得很慢，方向也错了，所以他坦率地指出了这一点。他有一个具体的建议：到地下去。当他因吃受火灾影响的猎物而遭到谴责时，他迅速为自己的行为辩护，并指出其他动植物，比如游隼和桉树，也会造成伤害。巨蜥有兴趣向非洲学习，并承认我们都应该向那些经历过苦难的人学习。他受到启发并意识到未来建筑应该建在地下，而不是建在更高的地面上。

树袋熊

树袋熊承受着巨大的压力，但仍具有反思的能力。她有自知之明，意识到自己的局限性。她很肯定地指责巨蜥的捕猎行为。她表明，尽管她可能动作缓慢，但她头脑敏捷且善于观察。她担心大火会蔓延到树梢。她了解大局，也了解热浪和干旱的作用。她分享了关于非洲的知识。她通过共享有关生存在地下的树木的信息来支持居住在地下的想法。树袋熊成功说服并启发了巨蜥。

艺术
The Arts

我们一直在地面上建造建筑。如今我们面临着全球变暖的现状，经历了更频繁也更惨烈的火灾。也许我们是时候开始建造地下建筑以便在那里安全地生活。自然界中成百上千的物种就是这样做的。一些国家已经考虑建立地下城市来保护他们的公民免受核攻击。古老的地下城市包括佩特拉（约旦）、奥尔维耶托（意大利）、拉利贝拉（埃塞俄比亚）和德林库尤（土耳其）。看看这些例子并设想你所生活的地下城市。现在画一些图来帮助你建构你的创意构想。

思维拓展
Systems: Making the Connections

随着温度升高，发生火灾的风险也在增加。一种适应火灾的生态系统在自然中已经建立好了，稀树草原生态系统就是一个很好的例子。动植物为了利用自然提供的机遇或者减轻生态系统变化带来的不利影响正在快速进化。人类需要适应自身已经造成和仍在造成的变化。正是在这样的背景下，非洲的地下森林代表了一种奇思妙想，这里的稀树草原为更多的野生动物提供了栖息地。人类住处（往往是不自知地）被安置在火灾易发地区。除了有火灾风险外，人们在进行城市规划时也没有重视当地环境的特点。因此，虽然气候变化是导致森林火灾增加的一个主要因素，但森林管理不善和居住区规划不当也导致了生命和财产损失。我们试图寻找保护自身和财产安全的方法，其中一种做法是使用阻燃剂，但这么做会传播有毒物质。我们需要一种新的根本性解决方法，比如，重新考虑人类居住区的位置。植物、动物、真菌和细菌已经尝试过这种选择。我们要向这些生物学习。一些现代和古代文明也已经树立了榜样，冒险建造了巨大的地下建筑从而让人们生活在相对安全的环境中。地下不仅更凉爽，而且控制温度也更容易。我们可以从那些成功适应这种更安全的环境的物种身上学习。仔细看看其他物种是如何在大自然的启发下生存进化，然后轮到我们在这个充满挑战与变革的时代探寻新机遇。

动手能力
Capacity to Implement

重要的是要知道如何灭火。最有可能发生的火灾是家中的电气火灾。水能导电，因此发生电气火灾时，不能用水将其扑灭。如果可以安全进行的话，第一步是拔下电线。如果你确切知道建筑物的主断路器在哪里，请赶紧用它切断电路。然后，使用粉状物质（如碳酸氢钠）灭火。当然，最好的办法是在房屋内安装一个专业的碳酸氢钠灭火器。制定火灾预案并让所有人接受培训，以便让他们知道火灾发生时该如何应对。

故事灵感来自
This Fable Is Inspired by

雷切尔·苏斯曼
Rachel Sussman

雷切尔·苏斯曼出生于1975年，是一名当代艺术家。1998年，她获得了美国纽约曼哈顿视觉艺术学院的摄影学位。2008年，她又在纽约州巴德学院获得硕士学位。屋久岛是联合国教科文组织认定的世界遗产，在参观了岛上数千年的树之后，她花了10年的时间来寻找各大洲现存最古老的树种。她致力于对时间的深刻理解。她发现了一些细菌、真菌和苔藓。在南非，她看到了只有树冠可见的巨大地下森林。她以写作、会议和艺术创作的方式宣传她的发现，包括那本由芝加哥大学出版社出版的著作《世界上最古老的生物》（2014）。

图书在版编目（CIP）数据

冈特生态童书.第七辑:全36册:汉英对照 /
(比)冈特·鲍利著;(哥伦)凯瑟琳娜·巴赫绘;
何家振等译.—上海:上海远东出版社,2020
ISBN 978-7-5476-1671-0

Ⅰ.①冈… Ⅱ.①冈… ②凯… ③何… Ⅲ.①生态
环境–环境保护–儿童读物—汉英 Ⅳ.①X171.1-49

中国版本图书馆CIP数据核字(2020)第236911号

策　　划　张　蓉
责任编辑　祁东城
封面设计　魏　米　李　廉

冈 特 生 态 童 书
地下生活真快乐
[比]冈特·鲍利　著
[哥伦]凯瑟琳娜·巴赫　绘

闫宇昕　译